Contents

		PAGE
Introduction		2
Amendments to the Approved Documents 1985, 1990 and 1992 Editions.		
A - Structure.	1992 Edition.	3
B - Fire Safety.	1992 Edition.	5
C - Site preparation and resistance to moisture.	1992 Edition.	9
D - Toxic substances.	1985 Edition.	11
E - Resistance to the passage of sound.	1992 Edition.	13
F - Ventilation.	1990 Edition.	15
G - Hygiene.	1992 Edition.	17
H - Drainage and waste disposal.	1990 Edition.	19
J - Heat producing appliances.	1990 Edition.	23
K - Stairs, ramps and guards.	1992 Edition.	25
L - Conservation of fuel and power.	1990 Edition.	27
M - Access and facilities for disabled people	1992 Edition.	29
N - Glazing - materials and protection.	1992 Edition.	31
Regulation 7- Materials and workmanship.	1992 Edition.	33

Note that the Manual to the Building Regulations 1985 has been withdrawn.

Approved Document   Amendments 1992.

Introduction

690.2
BOI

1 The publication of the 1992 Edition Approved Documents, Parts A,B,C,E,G,K,M,N, and Regulation 7, together with the 1990 Edition Approved Documents Parts F,H,J and L and the 1985 Edition of Part D Approved Document has now completed the Department's Stage Two review of the technical content of the Regulations.

2 This document includes corrections, clarifications and revisions which will bring the Approved Documents issued as 1985. 1990 and 1992 Editions to the Building Regulations up to date.

3 It should be noted that as a result of consultation the guidance in Part F regarding the provision of extract ventilation and room ventilated heating appliances has been qualified.

4 Following the issue of this document, subsequent reprints of the Approved Documents will include the amendments.

5 It is proposed to issue Amendments documents such as this in order to keep Approved Documents up to date.

6 Reference to Codes and Standards may be deleted if the particular Code or Standard is withdrawn by the British Standards Institution. However the withdrawn Standard may still provide guidance where it addresses the relevant requirements of the Regulations.

Approved Document A - Structure

Page 1 Last line, Standards referred to.
Amend Page no. to "68"

Page 4 Para 2.
Amend to read "DD ENV
1992-1-1:1992 Eurocode 2 :Part 1 and DD
ENV 1993-1-1:1992 Eurocode 3 :Part 1.1
General. . . . . . "

Page 10 Para 1B5. Line 4.
Delete "Part 1:"

Page 14 Diagram 3.
Lower diamond. Delete "building" substitute "wall".

Page 15 Table 6. Column 1. Line 2.
Substitute "76 - 100" for "5 - 100".

Page 16 Para IC12. Line 5.
Delete "and IC40".

Page 18 Table 7. Column 1. Line 2.
Add "floors".

Page 21 Diagram 11.
Key. Reference for height A1 insert "floor" between "ground" and "is".

Page 23 Diagram 13.
Delete Diagram 13 and insert the amended diagram below.

Diagram 13 Maximum span of floors

See para 1C24

a. FLOOR MEMBER BEARING ON WALL — wall, floor, floor span maximum 6m, centre line of bearing

b. FLOOR MEMBER BEARING ON JOIST HANGER — wall, floor, floor span maximum 6m, centre line of bearing

Page 24 Diagram 15.
Note 3. Delete "C.11" substitute "11".

Page 31 Para. 1E2. Design Provision
In d. amend referenced diagram to "Diagram 24a. and b".

In e. amend referenced diagram to "Diagram 22".

In f. amend referenced diagram to "Diagram 22".

In g. amend referenced diagram to "Diagram 23".

Diagram 22.
Delete "See para 1Ee. and f." and insert "See para 1E2e. and f".

Page 33 Para. 2.10. Line 2.
Insert "assumed" between "their" and "safe".

Line 3. Delete "loads" substitute "strength".

Para. 2.11. Line 2.
Insert "assumed" between "their" and "safe".

Delete "loads" substitute "strength".

Page 34 Para. 2.15. Line 2.
Delete "BS 5268" and insert "BS 5628".

Page 36 Para 4.6 Structural work of steel.
Amend date of BS. 5950 Part 2 to "1992"

Page 44 Table A1.
Amend the heading "Spacing of joist (mm)" to "Spacing of joists (mm)".

Page 45 Amend the heading "Spacing of joist (mm)" to "Spacing of joists (mm)".

Page 46 Table A3
Amend the heading "Spacing of joist" to "Spacing of joists (mm)".

Page 47 Table A4.
Amend the heading "Size of joist" to read "Size of binder".
Amend the heading "Spacing of joist (mm)" to read "Spacing of binders (mm)".

Page 48 Table A5.
Amend the heading "Size of joist" to read "Size of rafter".
Amend the heading "Spacing of joist (mm)" to read "Spacing of rafters (mm)".

Page 50 Table A7. ditto.

Page 52 Table A9. ditto.

Page 54 Table A11. ditto.

Page 56 Table A13. ditto.

Page 58 Table A15. ditto.

Page 53 Table A10. SC4/0.75.
For 50 x 200 timber purlins amend the "2.97" clear span given in the 12th column to "1.97"

Page 54 Table A11. SC3/1.0.
For 47 x 150 timber rafters amend the "2.46" clear span given in the 2nd column to "3.46".

Approved Document
Amendments 1992.

Page 49
Amend the heading "Size of joist" to read "Size of purlin".
Amend "spacing of joist (mm)" to read "spacing of purlins (mm)".

Page 51 ditto.

Page 53 ditto.

Page 55 ditto.

Page 57 ditto.

Page 59 ditto.

Page 60 Table A17.
Amend the heading "Spacing of joist (mm)" to read "Spacing of joists (mm)".

Page 61 Table A18. ditto.

Page 62 Table A19. ditto.

Page 63 Table A20. ditto.

Page 64 Table A21. ditto.

Page 65 Table A22. ditto.

Page 66 Table A23.
Amend the heading "Size of joist" to read "Size of purlin".

Amend the heading "Spacing of joist (mm)" to read "Spacing of purlins (mm)".

Amend "1.75" in the heading to the 4th column to read "0.75"

Table labelled "SC3/0.75" for "50 x 175" timber purlin 7th column. Delete "2.9" substitute "2.90".

Page 67 Table A24.
Amend the heading "Size of joist" to read "Size of purlin"

Amend the heading "Spacing of joist (mm)" to read "Spacing of purlins (mm)".

4th column. heading.
Delete "1.25" substitute "0.75".

Page 68 Add "68" to bottom of page.
Amend date of BS 5950 Part 2. to "1992"

Amend date of BS 5328 Part 1 to "1991"

Amend date of BS 5328 Part 2 to "1991"

Approved Document B - Fire safety

B1

Page 12 Headings "Security" and "Use of the document" should be the same type face as the other headings on the page.

Page 14 Para 1.5.
Replace the last sentence with:
"The smoke alarms may be wholly mains operated or mains operated with a secondary power supply such as batteries. Smoke alarms operated by primary batteries are not acceptable (see note to clause 23, BS 5446: Part 1: 1990)."

Para 1.17 Line 13.
Delete full stop, substitute comma.

Para 1.18(a) Line 4.
Delete "900mm" substitute "800mm".

Page 18 Para 1.30.
Add "In a two room loft conversion, a single escape window can be accepted provided both rooms have their own access to the stairs. A communicating door between the rooms must be provided so that it is possible to gain access to the escape window without passing through the stair enclosure."

Diagram 4. Dormer window.
Amend cill minimum height from "900mm" to "800mm".

Page 22 Para 2.17.
Replace colon with fullstop at the end of the paragraph.

Page 23 Para 1.
Add after first sentence "They are not applicable where the top floor is not more than 4.5m above ground level."

Para 2.19(b)
Amend "4" to "3".

Diagram 10 Key.
Add to automatic opening vent "(1.5m^2 minimum free area)". In diagram b the AOV should be positioned in the external wall of the common lobby instead of the stairway.

Page 24 Diagram 11. Key.
Add to automatic opening vent "(1.5m^2 minimum free area)".

Diagram c. Position "AOV" at left hand end of the corridor.

Page 26 Para 2.29. Line 1.
Delete "2.17" substitute "2.18"

Line 2. Delete "2.18" substitute "2.19".

Diagram 13 Key.
Add tone between lines representing fire resisting construction.

Page 29 Section heading omitted.
Add DESIGN FOR HORIZONTAL ESCAPE - BUILDINGS OTHER THAN DWELLINGS.

Page 32 Para 3.9(d) Line 3.
Insert "appropriate" before "limit"; delete "for". Delete line 4.

Page 34 Diagram 19.
Change the sign "≡" to "is equivalent to"

Page 37 Para 4.16(b).
Amend to "all stairs serving buildings with open spatial planning; and"

Worked example:
Amend to "from the formula:
1200 = 200w + 50(w - 0.3) (12 - 1)
1200 = 200w + 550w - 165
1365 = 750w
w = 1.82m"

In last paragraph amend "1.815m" to "1.82m".

Page 43 Para 5.33
Add to end of first para: "which illuminates the route if the mains supply fails".

In the second para. Line 2
Delete "protected" and "(see paragraph 5.35)".

Para 5.35. Heading.
Amend to "Protected power circuits."

Line 5 omit "lighting or".

B2

Page 45 Requirement B2(1)
Omit comma after "building".

Page 48 Para 6.12.
Amend last two lines to "tested as a part of the ceiling system that is to be used to provide the appropriate fire protection."

B3

Page 51 Para 0.47.
Penultimate para, third line from the bottom amend "0.43a" to "0.47a".

Page 61 Para 9.6. Line 3.
Change "15" to "16".

Page 62 Table 13. Item 9.
Change "15m" to "20m".

Page 63 Table 14. Title.
Amend to ". . . . non-domestic buildings (purpose groups 2-7)"

B4

Page 72 Para 12.7.
First para, penultimate line. Delete "Table 16," add "Diagram 36".

Second para, First line. Change "15m" to "20m".
Add new third paragraph:
"Advice on the use of thermal insulation material is given in the Building Research Establishment Report "Fire performance of external thermal insulation for walls of multi-storey buildings (BR 135, 1988)."

Page 80 Para 14.6. Line 3.
Amend to "unplasticized PVC".

Page 81 Table 17.
Replace table with:

Table 17 Limitations on roof coverings*

Designation of covering of roof or part of roof	Minimum distance from any point on relevant boundary			
	Less than 6m	At least 6m	At least 12m	At least 20m
AA, AB, or AC	●	●	●	●
BA, BB, or BC	○	●	●	●
CA, CB, or CC	○	●(1)	●(2)	●
AD, BD, or CD	○	●(1)	●(2)	●(2)
DA, DB, DC, or DD	○	○	○	●(1)
Thatch or wood shingles, if performance under BS 476: pt 3: 1958 cannot be established	○	●(1)	●(2)	●(2)

Notes:
Separation distance considerations do not apply to roofs of a pair of semi-detached houses (see 14.4)

* See paragraph 14.7 for limitations on glass, and paragraphs 14.5 and 14.6 and Table 18 for limitations on plastics rooflights.

● Acceptable
○ Not acceptable

(1) Not acceptable on any of the following buildings:
 a. Houses in terraces of three or more houses,
 b. Industrial, Storage or Other non-residential purpose group buildings of any size,
 c. Any other buildings with a cubic capacity of more than 1500m^3

And only acceptable on other buildings if the part of the roof is no more than 3m^2 in area and is at least 1.5m from any similar part, with the roof between the parts covered with a material of limited combustability.

(2) Not acceptable on any of the buildings listed under a. b. or c. above.

Amendments 1992.

Page 81 Table 18.
Replace table with:

Table 18 Plastics rooflights: limitations on use and boundary distance				
Classification on lower surface (1)	Space which rooflight can serve	Minimum distance from any point on relevant boundary to rooflight with an external surface classification(2) of:		
		TP(a)	AD BD CA CB CC CD OR TP(b)	DA DB DC DD
1. TP(a) rigid	any space except a protected stairway	6m(3)	6m(5)	20m
2. Class 3 or TP(b)	a. balcony, verandah, carport, covered way or loading bay, which has at least one longer side wholly or permanently open	6m	6m	20m
	b. detached swimming pool			
	c. conservatory, garage or outbuilding, with a maximum floor area of 40m²			
	d. circulation space(4) (except a protected stairway)	6m(5)	6m(5)	20m(5)
	e. room(4)			

Notes:
na Not applicable
(1) See also the guidance to B2
(2) The classification of external roof surfaces is explained in Appendix A.
(3) No limit in the case of any space described in 2a. b. and c.
(4) Single skin rooflight only, in the case of non-thermoplastic material.

(5) The rooflight should also meet the provisions of Diagram 42.
Polycarbonate and PVC rooflights which achieve a Class 1 rating by test, see paragraph 14.6, may be regarded as having an AA designation.
None of the above designations are suitable for protected stairways - see paragraph 6.10.
Products may have upper and lower surfaces with different properties if they have double skins or are laminates of different materials.

B5

Page 84 Para 15.5.
Amend "17.7 and 17.8" to "17.9 and 17.10".

Page 85 Para 16.2. Line 6.
After "building" insert 'or to 15% of the perimeter whichever is the less onerous'.

Page 87 Table 20. Column 4.
Turning circle between kerbs, amend "29.0" to "26.0".

Page 89 Para 17.3. Line 1.
After "Buildings" insert "other than open sided car parks (see 11.4)".

Table 21. First column.
Change heading to "Largest qualifying floor area (m²)".

Page 90 Para 17.7.
Add new para "c. the same 900m² per firefighting shaft criterion should be applied to calculate the number of shafts needed where basements require them."

Appendix A

Page 98 Table A2. Sub-heading immediately below "Ground or upper storey" change "separating" to "separated".

In row 1b and c, add "@" against "60".

In footnote, add "@ 30 minutes in the case of three storey dwelling houses increased to 60 minutes minimum for compartment walls separating buildings".

Appendix E

Page 114 Para: Separated part.
Amend "Diagram C3" to C4".

Appendix F

Page 118 in B1 and B3.
Delete "Fire Prevention Guide No.1. Fire . . . . redevelopment (Home Office), 1972".

Amend existing entry in B1 "Firecode HTM 88. Guide to Fire Precautions in NHS housing in the Community for mentally handicapped (or mentally ill) people".

Approved Document   Amendments 1992.

Index

Page 123 Index, amend references in:-
Fire fighting lifts,
Fire fighting lobbies,
Fire fighting shafts,
Fire fighting stairs,
Delete "diagram 47" add "diagram 46".

Approved Document C. Site preparation and resistance to moisture.

Page 13 Para 3.5. Line 2.
After "(1200 gauge)" add "250µm (1000 gauge) polyethylene in accordance with appropriate BBA certificate or to the PIFA standard is also satisfactory".

Page 14 Para 3.10 ii. Line 4.
After "(1200 gauge)" add "250µm (1000 gauge) polyethylene in accordance with an appropriate BBA certificate or to the PIFA standard is also satisfactory".

Page 22 Standards referred to.
BS 5618 add "AMD.6262".

Approved Document D - Toxic substances

Cover
Delete "Building Regulations 1985" add "Building Regulations 1991".

Inside front cover
Delete the text and insert as in Approved Documents, 1992 Edition.

Page 1 Line 3.
Delete "1985" add "1991".

Page 3 Delete "British" in title.
Standards referred to.
BS 5618: 1985. Add "AMD 6262".
BS 8208: Part 1: 1985. Add "AMD 4996".

Approved Document E - Resistance to the passage of sound

Page 3 Note after Regulation 6.
Clarification - Where a material change of use occurs and a dwelling is formed adjacent to or within another building type the requirement will apply.

Page 7 Para 1.3. Line 1.
After "habitable room" add "or kitchen".

Page 9 External wall. Line 25.
Delete "sealed" substitute "stopped".

Page 11 External wall.
Add after c. entry "Where the external wall has a cavity, the cavity should be stopped with a flexible closer."

Page 20 Floor type 2. F. Screed. Line 1.
Delete "55mm" and substitute "65mm".

Page 22 Floor type 3. Floors.
A. Platform floor with absorbent material.
Line 8.
Delete "80-100kg/m^3" add "60-100kg/m^3".
Line 10.
After "softer floor" add footnote "in such cases additional support can be provided around the perimeter of the floor by a timber batten with a foam strip along the top attached to the wall".
Penultimate line.
Delete "rock fibre" add "mineral wool".

C. Ribbed floor with heavy pugging.
Line 3.
After "nailed" insert "or screwed".

Page 23 D. Pugging. Line 3.
Insert comma after bracket.

Page 31 Floor treatment 2. Construction b.
Para 2. Line 3.
Delete "80 and 100kg/m^3" add "60-100kg/m^3". See footnote to Page 22 above.

Page 32 Floor treatment 3. Construction a. i.
Line 1.
Delete "of" substitute "or".

b. Para 3. Line 3.
Insert comma after bracket.

Page 35 Para 6.3. Line 7.
Delete "at least" substitute "not more than".

Approved Document F - Ventilation

Cover
Delete "The Building Regulations 1985" add "The Building Regulations 1991".

Page 2 Line 4.
Delete "1985" add "1991".

Page 3 After Paragraph 2.1.
Insert "**2.2. Mechanical extraction as set out in Paragraph 2.1.(i) with an open-flued appliance in the kitchen can cause the spillage of flue gases and could create dangerous conditions.**
In such situations it may be appropriate to reduce the provision set out in Paragraph 2.1(i).

2.3. A kitchen which is wholly internal and is ventilated as set out in Paragraph 2.1. above should not contain an open-flued appliance.

2.4. In other kitchens where there is an open-flued appliance and mechanical extraction is provided as set out in Paragraph 2.1.(i) then the appliance and flue should be able to operate effectively whether or not the fan is running.
For example with:-
i) Gas appliances a spillage test as described in BS 5440. Part 1. 1990. Clause 4.3.2.3. should be carried out.
ii) Oil-fired appliances, advice can be obtained from: Oil Firing Technical Association for the Petroleum Industry (OFTEC), Century House, 100 High Street, Banstead, Surrey. SM7 2NN.
iii) Solid fuel appliances - mechanical extraction should not be provided in the same room.

Page 5 Para 7.1. (c)
Delete "9.8 and" amend to "9.9 and 9.10".

Page 6 Line 4.
Delete "1985" add "1991"

Page 8 Para 1.5.
Delete "and 9.3" add "9.2 and 9.4".

Page 9 Para 2.8.
Delete "and 9.3" add "9.2 and 9.4".

Page 10 Standards referred to.
Amend "BS 5270" to "BS 5720"

Approved Document G - Hygiene

Inside front cover

MAIN CHANGES IN THE 1992 EDITION

G.1. b. Kitchen separation.
Last line.
Delete "1.3" and insert "1.2".

G.3. f.i.
Delete "temperature relief valve" and add "thermal cut-out".

G.3.

Page 7 Limits on application.
b). Delete comma after "system".

Page 12 Para 4.3.
Insert 'not' before "more than 45kW."

Page 13 Standards referred to.
BS 3955 add "AMD 5940"

Approved Document H - Drainage and waste disposal

Cover Delete "Building Regulations 1985" add "Building Regulations 1991"

H1

Page 2 Line 4.
Delete "1985" add "1991"

Page 3 Para 0.4. Last line.
Delete "8301" and substitute "5572".

Table 2.
Delete "(min dimension)" under diameter of trap for WC pan substitute "(siphonic only)".

Note at bottom of table.
Delete "40mm" and substitute "38mm".

Page 4 Diagram 1.
Add "min" to "200mm" radius at the foot of the centre illustration.

Add "min" to "200mm" on the right hand illustration.

Key.
Add "at least" after "offset".

Table 3. Column for "Max number to be connected".
Delete "6" and substitute "7".

Page 5 Diagram 3. Note.
Delete "lengthened by" and add to the end of the note "40mm washbasin waste pipes may slope between 18mm to 45mm/m."

Page 6 Para 1.21.
Add at the end "or appliances with integral traps (see paragraph 1.4)."

Para 1.24. Para 2.
Delete "and" and substitute a comma.
Add "siphonic" before "closets".
Add at the end of the sentence "and stacks serving washdown closets not less than 100mm."

Page 7 para 1.31. Line 3.
Add "Where necessary" before "Different".

Table 5.
Alongside "BS 416" add "BS 6087".

Para. 1.32. Line 4.
Delete "During this time".

Page 8 Para 2.6.
End of paragraph add "(see also paragraph 1.22)"

Page 9 para 2.14.
End of paragraph add "where necessary".

Table 7.
Delete "BS 85" substitute "BS 65, BSEN 295".

Alongside "BS 437" add ",BS 6087".

Page 12 Table 11. Plastics.
Delete "BBA Certificate" substitute "BS 7158".

Page 13 Table A2. Urinal stall.
Delete "6 person" substitute "7 person".

H2

Page 16 Line 4.
Delete "1985" add "1991".

Page 18 para 1.7. Line 5.
After "600mm" add "where entry is required".

H3

Page 19 Line 4.
Delete "1985" add "1991".

H4

Page 23 Line 4.
Delete "1985" add "1991".

Page 25 para 1.9.
Add heading "Alternative approach".

Standards referred to

H1

BS 65: 1991 Specification for vitrified clay pipes, fittings and ducts, also flexible mechanical joints for use solely with surface water pipes and fittings.

BSEN 295: Vitrified clay pipes and fittings and pipe joints for drains and sewers
Part 1: 1991 Test requirements
Part 2: 1991 Quality control and sampling
Part 3: 1991 Test methods.

BS 416 Discharge and ventilating pipes and fittings, sand-cast or spun in cast iron,

Part 1: 1990 Specification for spigot and socket systems.

Part 2: 1990 Specification for socketless systems."

BS 437: 1978 Specification for cast iron spigot and socket drain pipes and fittings.

Amendment slip number   1: AMD 5877.

BS 864 Capillary and compression tube fittings of copper and copper alloy.

Part 2: 1983 Specification for capillary and compression fittings for copper tubes.

Amendment slip number   1: AMD 5097
                        2: AMD 5651.

BS 882: 1983 Specification for aggregates from natural sources for concrete.
Amendment slip number   1: AMD 5150.

BS 2871 Specification for copper and copper alloys. Tubes.

Part 1: 1971 Copper tubes for water, gas and sanitation.
Amendment slip number   1: AMD 1422
                        2: AMD 2203.

BS 3656: 1981 (1990) Specification for asbestos-cement pipes, joints and fittings for sewerage and drainage.

Amendment slip number   1: AMD 5531.

BS 3868: 1973 (1980) Specification for prefabricated drainage stack units: galvanized steel.

BS 3921: 1985 Specification for clay bricks.

BS 3943: 1979 (1988) Specification for plastics waste traps.
Amendment slip number   1: AMD 3206
                        2: AMD 4191
                        3: AMD 4692.

BS 4514: 1983 Specification for unplasticized PVC soil and ventilating pipes, fittings and accessories.

Amendment slip number   1: AMD 4517
                        2: AMD 5584.

BS 4660: 1989 Specification for unplasticiszed polyvinyl chloride (PVC-U) pipes and plastics fittings of nominal sizes 110 and 160 for below ground drainage and sewerage.

BS 5254: 1976 Specification for polypropylene waste pipe and fittings (external diameter 34.6mm, 41.0mm and 54.1mm).
Amendment slip number   1: AMD 3588
                        2: AMD 4438.

BS 5255: 1989 Specification for thermoplastics waste pipe and fittings.

BS 5481: 1977 (1989) Specification for unplasticized PVC pipe and fittings for gravity sewers.
Amendment slip number   1: AMD 3631
                        2: AMD 4436.

BS 5572: 1978 Code of practice for sanitary pipework.
Amendment slip number   1: AMD 3613
                        2: AMD 4202.

BS 5911 Precast concrete pipes fittings and ancillary products.

Part 2: 1982 Specification for inspection chambers and street gullies.
Amendment slip number   1: AMD 5146,

Part 100: 1988 Specification. for unreinforced and reinforced pipes and fittings with flexible joints.

Amendment slip number   1: AMD 6269

Part 101: 1988 Specification for glass composite concrete (GCC) pipes and fittings with flexible joints.

Part 120: 1989 Specification for reinforced jacking pipes with flexible joints.

Part 200: 1989 Specification for unreinforced and reinforced manholes and soakaways of circular cross section.

BS 6087: 1990 Specification for flexible joints for grey or ductile cast iron drain pipes and fittings (BS 437) and for discharge and ventilating pipes and fittings (BS 416).

Amendment slip number: 1: AMD 6357.

BS 7158: 1989 Specification for plastics inspection chambers for drains.

BS 8110 Structural use of concrete.

Part 1: 1985 Code of practice for design and construction

Amendment slip number   1: AMD 5917
                        2. AMD 6276.

Approved Document

BS 8301: 1985 Code of practice for building drainage.
Amendment slip number   1: AMD 5904

H2

BS 5328: Concrete.

Part 1: 1991 Guide to specifying concrete.

Part 2: 1991 Methods for specifying concrete mixes.

Part 3: 1990 Specification for the procedures to be used in producing and transporting concrete.

Amendment slip number 1: AMD 6927.

Part 4: 1990 Specification for the procedures to be used in sampling, testing and assessing compliance of concrete.

Amendment slip number 1: AMD 6928.

BS 6297: 1983 Code of practice for design and installation of small sewage treatment works and cesspools.

H3

BS 6367: 1983 code of practice for drainage of roofs and paved areas.
Amendment slip number   1: AMD 4444.

BS 8301: 1985 Code of practice for building drainage.
Amendment slip number   1: AMD 5904.
                         2: AMD 6580.

H4

BS 5906: 1980 (1987) Code of practice for the storage and on-site treatment of solid waste from buildings.

Approved Document J - Heat producing appliances

Cover Delete "Building Regulations 1985" add "Building Regulations 1991".

Page 2 The Requirement. Line 4. Delete "1985" add "1991".

Page 3 Para 1.3.
Delete "then the appliance and flue.............fan is running." add "See amendments to Part F para 2.1 to 2.4."

Page 4 Table 2. Installation.
Delete "Inglenook recess appliances" and substitute "Fireplace recess with an opening in excess of 500mm x 550mm"

Delete third entry. "Open fire   200mm diameter or square section of equivalent area".

Page 5 Para 2.12. (a). Line 2.
Delete "1971 (1977)" add "1989".

Page 6 Para 2.16 (a). Line 3.
Delete "1976" add "1990".

Line 5.
Delete "1976" add "1990" and after "chimneys" delete "for" and add "with stainless steel flue linings for use with".

Page 8 Para 3.1. Line 7.
Delete "BS 6714: 1986 Specification for installation..... appliances (1st, 2nd and 3rd family gases) or" and insert "BS 5871: Specification for installation of gas fires, convector heaters, fire/back boilers and decorative fuel effect gas appliances. Part 1: 1991 Gas fires, convector heaters and fire/back boilers (1st, 2nd and 3rd family gases).
Part 2: 1991 Inset live fuel effect gas fires of heat input not exceeding 15kW (2nd and 3rd family gases)
Part 3: 1991 Decorative fuel effect gas appliances of heat input not exceeding 15kW (2nd and 3rd family gases)".

Para 3.9. (b). Line 2.
Delete "1973 (1984)" and add "1989".

Line 4.
Delete "1973 (1984)" and add "1989".

(d). Line 1.
Delete "1" and substitute "2".

Page 9 3.11. (a). Line 2.
Delete "1971 (1977)" and add "1989".

In (b). Delete "1981" and substitute "1991".
In (c). Delete "1" and substitute "2".

Para 3.13. (a)
delete "1986" and add "1989".

Para 3.15.
After the first sentence delete the remainder and substitute "Any chimney that passes through, or forms part of a compartment wall or floor, must have walls that achieve the same degree of fire resistance required for that wall or floor (see Approved document B3)."

Page 10 Alternative approach.
BS 5546. Delete "1979 Code of practice" and substitute "1990 Specification".

BS 5864. Delete "1980 Code of practice" and substitute "1989 Specification".
BS 5871. Delete entry and substitute "BS 5871. Installation of gas fires, convector heaters, fire/back boilers and decorative fuel effect gas appliances.
Part 1: 1991
Part 2: 1991
Part 3: 1991"

BS 6172. Delete "1982 Code of practice" substitute "1990 Specification".

Delete "(2nd family gases)" and substitute "(1st, 2nd and 3rd family gases)".

BS 6173. Delete "1982 Code of practice" and substitute "1990 Specification".

Delete "(2nd family gases)" and substitute "(1st, 2nd and 3rd family gases)".

Add to list

"BS 5440: Installation of flues and ventilation for gas appliances of rated input not exceeding 60kW (1st, 2nd and 3rd family gases). Part 1: 1990 Specification for installation of flues.

Part 2: 1989 Specification for installation of ventilation for gas appliances".

Page 11 Para 4.8. (a).
For BS 4543 Amend to: "BS 4543 Factory-made insulated chimneys
Part 1: 1990. Methods of Test
Part 3: 1990 Specification for chimneys with stainless steel flue lining for use with oil-fired appliances."

Standards referred to

J 1/2/3

BS 41: 1973 (1981) Specification for cast iron spigot and socket flue or smoke pipes and fittings.

BS 65: 1991 Specification for vitrified clay pipes, fittings and ducts also flexible mechanical joints for use solely with surface water pipes and fittings.

BS 476 Fire tests on building materials and structures, Part 4: 1970 (1984). Non-combustibility tests for materials. AMD 2483 AMD 4390,

BS 567: 1973 (1989) Specification for asbestos-cement flue pipes and fittings, light quality.
AMD 5963

BS 715: 1989 Specification for metal flue pipes, fittings, terminals and accessories for gas-fired appliances with a rated input not exceeding 60 kW.
AMD 6615
AMD 6335

BS 835: 1973 (1989) Specification for asbestos-cement flue pipes and fittings, heavy quality.
AMD 5964

BS 1181: 1989 Specification for clay flue linings and terminals.

BS 1289: Flue blocks and masonry terminals for gas appliances, Part 1: 1986. Specification for precast concrete flue blocks and terminals; Part 2: 1989 Specification for clay flue blocks and terminals.

BS 1449: Steel plate, sheet and strip, Part 2: 1983 Specification for stainless and heat resisting steel plate, sheet and strip. AMD 4807 AMD 6646.

BS 4543 Factory-made insulated chimneys, Part 1: 1990. Methods of test. Part 2: 1990. Specification for chimneys with stainless steel flue linings for use with solid fuel fired appliances.
Part 3: 1990 specification for chimneys with stainless steel flue linings for use with oil fired appliances.

BS 5258 Safety of domestic gas appliances, Part 1: 1986. Specification for central heating boilers and circulators; Part 4: 1987 Specification for fanned-circulation ducted-air heaters; Part 5: 1989 Specification for gas fires; Part 7: 1977 Storage water heaters; Part 8: 1980 Combined appliances: gas fire/back boiler; Part 12: 1990 Specification for decorative fuel effect gas appliances (2nd and 3rd family gases); Part 13: 1986 Specification for convector heaters.

BS 5386 Specification for gas burning appliances, Part 1: 1976 Gas burning appliances for instantaneous production of hot water for domestic use, AMD 2990, AMD 5832; Part 2: 1981 (1986) Mini water heaters (2nd and 3rd family gases); Part 3: 1980 Domestic cooking appliances burning gas, AMD 4162, AMD 4405, AMD 4878, AMD 5220 and AMD 6642; Part 4: 1991 Built-in domestic cooking appliances.

BS 5410 Code of practice for oil firing, Part 1: 1977 Installations up to 44kW output capacity for space heating and hot water supply purposes, AMD 3637.

BS 5440 Installation of flues and ventilation for gas appliances of rated input not exceeding 60kW (1st, 2nd and 3rd family gases).
Part 1: 1990 Specification for installation of flues.
Part 2: 1989 Specification for installation of ventilation for gas appliances.

BS 5546: 1990 Specification for installation of gas hot water supplies for domestic purposes (1st, 2nd and 3rd family gases).
AMD 6656.

BS 5864: 1989 Specification for installation in domestic premises of gas-fired ducted air heaters of rated output not exceeding 60kW.

BS 5871: 1980 (1983) Specification for installation of gas fires, convector heaters fire/back boilers and decorative fuel effect gas appliances

Part 3: 1991 Decorative fuel effect gas appliances of heat input not exceeding 15kW (2nd and 3rd family gases).
AMD 7033.

BS 6172: 1990 Specification for installation of domestic gas cooking appliances (1st, 2nd and 3rd family gases).

BS 6173: 1990 Specification for installation of gas fired catering appliances for use in all types of catering establishments. (1st, 2nd and 3rd family gases).

BS 6461 Installation of chimneys and flues for domestic appliances burning solid fuel (including wood and peat), Part 2: 1984 Code of practice for factory-made insulated chimneys for internal applications.

BS 6798: 1987 Specification for installation of gas-fired hot water boilers of rated input not exceeding 60kW.

BS 6999: 1989 Specification for vitreous enamelled low-carbon-steel flue pipes, other components and accessories for solid fuel burning appliances with a maximum rated output of 45kW.

BS 8303: 1986 Code of practice for installation of domestic heating and cooking appliances burning solid mineral fuels, AMD 5723.

Amendments 1992.

Approved Document

Approved Document K - Stairs, ramps and guards

Page 3 Limits on application. Line 2.
Insert "only" after "apply".

Page 7 Para 1.21. Line 5.
Delete "1.1-1.5" and substitute "1.1, 1.2, 1.4 and 1.5".

Page 8 Para 1.30.
Delete the last sentence "Where glazing......; glazing materials and protection."

Page 10 Limits on application.
Delete paragraphs 1 and 2.

Page 11 Diagram 11. Assembly
Delete "within" before "530mm".
Delete figures for Strength and add note- "refer to BS 6399 Part 1"

All buildings.
Add note "except roof windows in loft extensions, see Approved document B1. Diagram 4".

Page 12 Para 3.5.
Delete "BS 6180: 1982 Code of practice for protective barriers in and about buildings and"
Diagram 13.
Delete "-3.7" substitute "and 3.5".

Diagram 14.
For para delete "3.6" substitute "3.5".

Approved Document
Amendments 1992.

Approved Document L - Conservation of fuel and power

Cover
Delete "Building Regulations 1985" add "Building Regulations 1991"

Page 2 Line 4.
Delete "1985" add "1991".
Contents. Section 3.
Delete "Commercial and Industrial installations 10".

Page 5 Para 1.3.
Title to read "Simple alternatives for dwellings."
Diagram 2.
Add "1" to Note.
Add "Note 2. BRE Information Paper IP3/90 provides an alternative method for determining insulation requirements for floors."

Page 8 Para 2.2
Add at the end of the paragraph "Night storage heaters are excluded."

Para 2.3.
Between paragraphs (a) and (b) add "or".

Page 10
Delete title "COMMERCIAL AND INDUSTRIAL INSTALLATIONS".
Para. 3.3. (b)
Delete "BS 5422: 1977" and substitute "BS 5422: 1990".

Page 12 Table 3.
Heading to col. 2 should be "Density (kg/m^3)"

Phenolic Foam thermal conductivity.
Delete "0.04" and substitute "0.02".

Notes. Line 3.
Delete "required" substitute "recommended".
Line 4
Delete "1980" substitute "1986".
Delete text in Section A3 and substitute the text in the second column of this page.

Page 14 Example 1. Third line from end.
Delete "0.75m^2/KW" substitute "0.75m^{2}K/W".

CALCULATION OF U VALUE

A3 This is best explained by an actual calculation. Consider the following exposed wall:

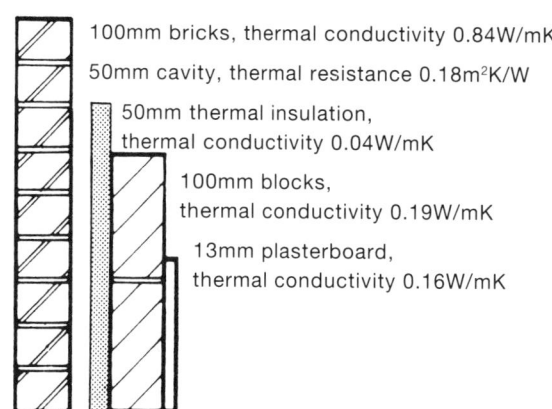

100mm bricks, thermal conductivity 0.84W/mK
50mm cavity, thermal resistance 0.18m^{2}K/W
50mm thermal insulation, thermal conductivity 0.04W/mK
100mm blocks, thermal conductivity 0.19W/mK
13mm plasterboard, thermal conductivity 0.16W/mK

In addition, resistance of outside surface of wall is 0.06m^{2}K/W and of inside surface 0.12m^{2}K/W.

The combined resistance of the construction is determined as follows:

	m^{2}K/W
resistance of outside surface	= 0.06
resistance of brick leaf = $\frac{0.1}{0.84}$	= 0.12
resistance of cavity	= 0.18
resistance of insulation material = $\frac{0.05}{0.04}$	= 1.25
resistance of block leaf = $\frac{0.1}{0.19}$	= 0.53
resistance of plasterboard = $\frac{0.013}{0.16}$	= 0.08
resistance of inside surface	= 0.12
total resistance (m^{2}K/W)	= 2.34

U value of construction = $\frac{1}{2.34}$ = **0.43**W/m^{2}K

Page 18 Table 6. Column (6).
Incorrectly shows ● no insulation required, delete "●" add "8" on first line and "10" on second line.

Note at the end of the page
Delete "there is no simple calculation available as an alternative to using Table 6." add "See BRE Information Paper IP3/90 for an alternative approach."

Example 7. Second line from bottom Col.1
Delete "10 - 15mm" add "10 - 15m".

Page 19 Table 7. Column (6).
Incorrectly shows ● no insulation required, delete "●" add "7" on first, line "8" on second line and "9" on third line.

Note at the end of the page
Delete "there is.....using table 7." add "See BRE Information Paper IP3/90 for an alternative approach."

Page 20 Example 9. Third line from end.
Delete "$1.11m^2/kW$" add "$1.11m^2K/W$".

Page 21 Para B12 (b).
Add "but noting the glazing constraint in paragraph 1.14 (page 6)".

Page 23 Standards referred to.
BS 699: 1984 (1990). Add "AMD 6600".

BS 1566: Part 1: 1984. Add "AMD 6598"
Part 2: 1984. Add "AMD 6601"

BS 3198: 1981. Add "AMD 6599".

BS 5422. Delete date and title and substitute "1990. Methods for specifying thermal insulation materials on pipes, ductwork and equipment in the temperature range of -40 to +700°C."

Approved Document M - Access and facilities for disabled people

Page 3 Requirement. M.1. Line 1.
Delete "part" substitute "Part".

M.2. Line 1.
Delete comma.

Page 6 Para. 1.1. Line 3.
Delete "building site" add "site curtilege".

Page 10 Para 1.31c.
Add "where" at the start of the paragraph.

Page 12 Diagram 8. Entrance lobbies.
Drawing top right 2000mm dimension should be between wall surfaces.

Page 13 Diagram 9. Internal doorways.
Delete para "1.35" add "2.4".

Para. 2.5. Line 6.
Delete "inaccessible by lift" add "to which lift access is not provided".

Page 16 Para 2.19. Line 7.
Delete "Practical reasons, such as the need for soft floor coverings, may militate against" add "For internal stairs it is not considered reasonable to require".

Para 2.20.
Amend to "If there is no lift access in a building, a stair suitable for people with walking difficulties should be provided. In any event, a stair should be suitable for people with impaired sight."

Diagram 12.
Delete para "1.21" add "2.21".

Delete landing height "1200mm" add "1800mm".

Title A to top drawing. Delete "EXTERNAL" add "INTERNAL"

Page 21 Diagram 16.
Plan dimension, omitted, should be "2000mm".

Toilet height. Delete "450mm to top of pan" add "450-475mm to top of seat".

Height to top of horizontal support at side of WC to remain at 700mm from finished floor level.

Diagram 17.
Toilet height. Delete "450mm to top of pan" add "450mm to top of seat".
Height to top of horizontal support at side of WC to remain at 700mm from finished floor level.

Approved Document N - Glazing-materials and protection

Page 3 Requirement N1.
Add comma after "Glazing".

Page 5 Para 1.1 a). Line 2.
After "internal" add "and external".
Para 1.2 c)
Delete "paragraph 1.7" add "paragraphs 1.7 and 1.8".

Para 1.3. Lines 1 and 2.
Delete "which, in practice, is concerned with the performance of laminated and toughened glass".
Add new paragraph to end "In terms of safe breakage, a glazing material suitable for installation in a critical location would satisfy the test requirements of Class C of BS 6206 or, if it is installed in a door or in a door side panel and has a pane width exceeding 900mm, the test requirements of Class B of the same standard."

para 1.4. Line 8.
Delete "area" substitute "dimension".

Diagram 2. Title.
Delete "thickness/area" add "thickness/dimension".

Page 6 Para 1.5.
Add to end of paragraph "traditional leaded lights or copper-lights."

Diagram 3
Note at bottom amend to "not to exceed 0.5m^2"

Para 1.6. Line 6.
Add to end of sentence "except in traditional leaded or copper-lights in which 4mm glass would be acceptable, when fire resistance was not a factor".

Page 7 Para 2.1. Line 6.
Delete "clear" add "transparent"
Para 2.2 Line 4.   ditto
Para 2.3 Line 2.   ditto
Para 2.6a) Line 1. ditto
    2.6b) Line 1. ditto
Diagram 5. Title   ditto

Para 2.1. Line 8.
After "walls" add "and doors".

Approved Document

Amendments 1992.

31